FUNDAMENTALS OF THE UNIFIED FIELD THEORY

KIKA

Amazon 2025

INTRODUCTION

For the past hundred years, physicists have unsuccessfully attempted to develop a unified field theory that explains all types of interactions in nature based on a single cause, using the scientific method of cognition.

This book attempts to lay the foundation for the development of a unified field theory (or "theory of everything") based on a scientific-philosophical approach [1]. This approach has made it possible to qualitatively describe the principles of the unified field theory. Even without quantitative analysis, this description already provides a coherent explanation for all known types of interactions. Moreover, the proposed foundations help resolve many unresolved problems in physics, such as the nature of dark energy, dark matter, empty spaces in the cosmos, galaxies without dark matter, antimatter, and other phenomena.

I am a mechanical engineer by education and hold a Ph.D. in technical sciences. I have worked as a design engineer, research associate, and programmer. Over the years, I have held positions as a team leader, laboratory head, and deputy head of a computing center. I have developed automated control systems for factories and research institutes. From 1995 to 2022, I worked for an American company, focusing on the creation, enhancement, and maintenance of a macro system designed to improve programmers' productivity and work quality. I am now retired.

For readers interested in fundamental worldview questions, this book will undoubtedly be valuable. It presents ideas that significantly differ from those of modern physics and offers explanations for phenomena that remain unresolved by current scientific theories.

I would like to express my gratitude to Timothy Kevorkov for the illustrations in this book.

TABLE OF CONTENTS

PREFACE

Unified field theories (or "theories of everything") are physical theories that aim to provide a single description of all known physical phenomena based on a unified primary field.

Historically, both classical and quantum theories have been proposed [2]. Classical theories include, for example, M-theory, while quantum theories encompass quantum gravity and string theory. There have also been attempts to create incomplete unified field theories that incorporate only two or three of the four known interactions. Such theories include generalized gravity theory, quantum field theory, the Standard Model, and Grand Unified Theories.

From the mid-1980s to the mid-1990s, string theory experienced rapid development. Many anticipated that it would soon lead to the formulation of the so-called "unified theory" or "theory of everything"—a goal that Einstein pursued unsuccessfully for decades. However, despite the mathematical rigor and internal consistency of string theory, no experimental confirmation has yet been found [3].

The theory of everything is a hypothetical unified physical-mathematical framework that describes all known fundamental interactions. Throughout the 20th century, numerous "theories of everything" were proposed, but none have successfully passed experimental validation [4].

Thus, despite extensive efforts by many physicists, a unified field theory has yet to be realized. The reason for these failures can be explained as follows: just as one cannot see an entire house from inside it, one cannot solve a global problem while remaining confined within a single domain. Einstein himself recognized this, stating that "problems cannot be solved at the same level of thinking at which they were created."

Therefore, when addressing fundamental challenges such as the development of a unified field theory, it is necessary to move beyond conventional scientific methods and adopt a scientific-philosophical approach. This approach combines the strengths of both scientific and philosophical methods while avoiding their limitations. But what exactly is the scientific-philosophical form of knowledge?

CHAPTER 1. SCIENTIFIC AND PHILOSOPHICAL FORM OF KNOWLEDGE

The scientific form of knowledge is a process of obtaining objective knowledge aimed at identifying the fundamental laws of the material world. Scientific knowledge operates on two levels: empirical and theoretical, both of which rely exclusively on experimental data.

The advantage of the scientific approach lies in its objectivity—it allows for the acquisition of knowledge that accurately reflects the laws of nature and can be practically applied. Thanks to scientific knowledge, modern technology and advanced engineering have been developed.

However, science also has its limitations. It is based on experimental data, which, despite rigorous verification, does not always correspond to absolute truth. What are facts? Facts are pieces of information obtained through our senses or scientific instruments.

Firstly, our senses have inherent limitations that prevent us from fully perceiving reality. While scientific instruments extend our perceptual capabilities, they, too, have constraints, making the data they provide not always completely accurate. For example, space is perceived by scientific instruments as isotropic, whereas in reality, it is anisotropic (see the section

"Anisotropy of Space" in Chapter 5, "The Birth of Matter").

Another example: the actual trajectories of inertial motion may be wave-like, helical, or a complex combination of these (see "Trajectories of Inertial Motion" in the section "Motion" of Chapter 5, "The Birth of Matter"). However, due to the small diameters of the helical lines and the amplitudes of the wave lines, the instruments perceive them as straight. Many other such examples exist.

Secondly, the same set of facts can often be explained in multiple ways. Therefore, even widely accepted scientific explanations are not necessarily true. These two points indicate that scientific knowledge should be approached with caution, as its absolute truth is not guaranteed.

Philosophical knowledge, in contrast to scientific knowledge, does not focus on fragmented aspects of reality but seeks to understand the world as a whole. It is based on certain fundamental worldview principles and aims to uncover the foundational laws of existence and humanity's place within it.

The strength of philosophical inquiry lies in its reliance on immutable fundamental principles, which possess universal validity. However, its limitation is its lack of practical applicability, as it does not provide specific knowledge about fragmented aspects of reality.

Scientific disciplines independently study nature from different perspectives, leaving many interdisciplinary questions unresolved. Similarly, different forms of inquiry—scientific and philosophical—approach reality from distinct standpoints, leaving gaps in understanding.

To bridge these gaps, just as interdisciplinary scientific fields have emerged (e.g., chemical physics, biophysical chemistry), new hybrid forms of knowledge must also be developed. One such method is the scientific-philosophical method of knowledge.

A key example of an unresolved interdisciplinary issue is the unified field theory, proposed by Einstein nearly a century ago. This problem lies at the intersection of philosophy and physics and can only be effectively addressed through a scientific-philosophical approach.

The essence of this approach is as follows: like scientific knowledge, it seeks to uncover the fundamental laws of the material world. However, unlike science, it does not rely on experimental data but rather on fundamental principles governing the existence of matter. But what are these fundamental principles?

CHAPTER 2. FUNDAMENTAL PRINCIPLES OF THE EXISTENCE OF MATTER

The scientific-philosophical method of knowledge was first formulated during the development of the Hypothesis of Universal Interaction [5]. As part of this work, seven fundamental principles of material existence were identified:

1. **The Principle of Finiteness.** This principle states that our Metagalaxy has finite dimensions, even though it may appear infinite to us. It also implies that all processes occurring in the world, including the world itself, have a limited lifespan.
2. **The Principle of Organization.** This principle asserts that structures exist at all levels of the material world, from the microscopic scale (elementary particles, atoms, molecules) to the macroscopic scale (planets, stars, galaxies). It confirms that chaos does not exist in nature.
3. **The Principle of Discreteness (Quantization).** This principle states that all levels of matter organization (fields and substance), its forms of existence (space, motion, time), and its qualities (mass, inertia, friction) possess a quantum structure. This means that their smallest units (quanta) cannot be divided further without losing their essential properties. For example, substances are composed of molecules, but when molecules

break down into atoms, the original substance ceases to exist. Similarly, chemical elements lose their identity when atoms break down into elementary particles.

4. **The Principle of Interconnectivity.** This principle asserts that all objects are interconnected not only within the same level of matter organization (atoms and molecules, plants and animals, stars and planets) but also across different levels (celestial bodies and living organisms, elementary particles and galaxies).
5. **The Principle of Duality.** This principle recognizes the dual nature of all material objects. For instance, every object has not only a visible component but also an invisible field surrounding it. This applies to all objects without exception, including living beings.
6. **The Principle of Materiality.** This principle asserts that not only visible or instrumentally detectable objects and fields are material, but also all phenomena and processes in the world, even those that cannot yet be measured by modern scientific methods. These include life energy, emotions, and thoughts. This principle also implies that absolute emptiness does not exist.
7. **The Principle of Similarity.** This principle highlights the presence of structural similarities between objects at different levels of existence. For example, by studying the structure of the human body, one can infer the structure of the cosmos.

CHAPTER 3. MAN AND SPACE

Ancient sages observed the principle of similarity and expressed it in the words: **"As above, so below. As below, so above."** This principle allows us to construct a proper understanding of the cosmos by drawing an analogy with the structure of man. It is no coincidence that man is often called a **"microcosm."**

However, different philosophical schools offer widely varying perspectives on human nature, especially regarding the psyche. Due to these discrepancies, we will develop our understanding of a cosmos analogous to man based only on clear and indisputable facts.

MAN

The following can be said about a man.

First, a person can exist only in a specific environment that provides the ability to breathe, eat, move, and perform other essential life functions.

Second, in addition to a visible physical body, a person possesses an invisible psyche. Without it, he becomes a **"living corpse."**

Third, human actions are determined by thoughts and emotions, that is, by the psyche. Otherwise, even a **"living corpse"** would be capable of meaningful actions.

A human life cycle consists of seven stages:

- pre-embryonic state;
- embryo formation;
- prenatal development;
- birth;
- growth and development;
- aging;
- death.

SPACE

Now, using the principle of similarity, let us attempt to form a general understanding of the cosmos as something analogous to a human being.

According to this principle, the following must be acknowledged:

The cosmos, like a human, can exist only in a specific space (environment) that provides the conditions for its movement and development.

Just as a human possesses not only a physical body but also a psyche, the cosmos has an invisible component alongside its visible structure. For convenience, we will refer to this as the **cosmic energy field**. Without this field, the active cosmos would become passive—just as a human without a psyche turns into a **"living corpse."**

Everything that happens in the cosmos is determined by the influence of its energy field, just as human actions are predetermined by their psyche.

According to the principle of similarity, the cosmos (like a human) has seven stages in its life cycle:

- the state before the emergence of matter (**pra-matter**);
- the birth of matter;
- prenatal development (formation of particles);
- birth of cosmic bodies;
- growth and development of cosmic bodies;
- aging of cosmic bodies;
- the death of the cosmos.

Now, let us attempt to detail the stages of the cosmic life cycle.

CHAPTER 4. PRA-MATTER

Just like a human, the cosmos has a **pre-embryonic state**, which enables the emergence of matter.

What exactly is this pre-embryonic state?

ETHER

Just as a fetus needs the womb, the emerging cosmos requires a similar environment. Just as a newborn depends on its mother, especially in the early stages of life, this environment remains essential for the cosmos after its emergence. Therefore, it can rightfully be called **pra-matter**—that which preceded and gave rise to the cosmos (i.e., matter).

The name of this pra-matter is not of fundamental importance. In this work, we use the term introduced by ancient thinkers—**ether**. Pra-matter is ether in its original state, or **pra-ether**. According to the **principle of finiteness**, pra-ether is not infinite; it has boundaries. In accordance with the **principle of quantization**, it consists of the smallest portions—**quanta of ether**. Over time, pra-ether transforms into what we now recognize as the cosmos.

The mysterious **dark energy** [6], which constitutes 74% of the entire universe, is an expression of pra-ether. Additionally, pra-ether includes the vast empty regions of space stretching millions of light-years across. In 2007, American astronomers discovered the largest of these voids [7], calling it **the Great Nothing**, due to the absence of any matter within it.

What, then, is the **space of pra-ether**?

SPACE

According to the **principle of finiteness**, space is not infinite and has its boundaries. In accordance with the **principle of quantization**, space consists of the smallest units — **quanta of space**. Each quantum of space has its own content and shape.

The content of a quantum of space is the **quantum of ether**. The shape of the quantum of space could be the shape of the most perfect figure — the **sphere**. However, spheres cannot fill all of space without gaps. The presence of emptiness contradicts the **principle of materiality**. Therefore, the shape of the quantum of space is a polyhedron that minimally differs from a sphere but allows all space to be filled without voids. This polyhedron is a **rhombohedron** (see fig. 1).

In the orientation of the rhombohedron shown in fig. 1, three zones can be identified: the upper, middle, and lower zones. Each zone contains four faces. Faces of the middle zone are conditionally called northern, southern, eastern, and western. In fig. 1, only the southern and eastern faces of the middle zone are visible, and the south-western and south-eastern faces are visible in the upper and lower zones.

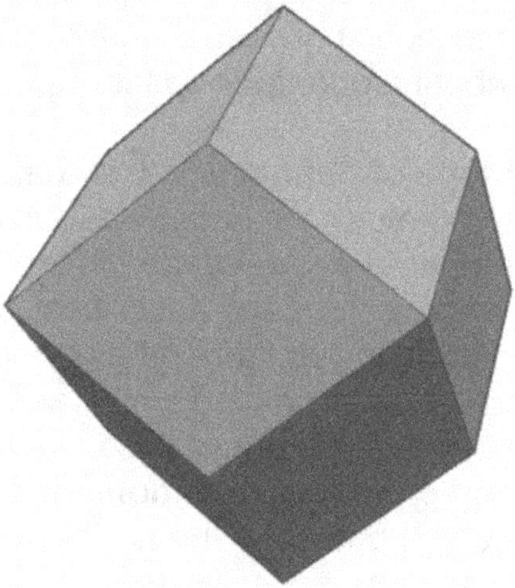

Fig. 1

According to the **principle of duality**, space consists of two interwoven half-spaces that are isolated from one another. These half-spaces can be conditionally called **positive** and **negative**. Both half-spaces are identical and together fill all of space without voids. They form a network of interconnected quanta of space.

Each quantum of space in this network is connected to four neighboring quanta of space: two in the upper zone and two in the lower zone. The middle zone has no connected quanta. For brevity, the word "neighboring" will refer specifically to "connected neighboring" quanta of space.

According to the principle of duality, there are two types of connections between a quantum of space and its neighboring quanta of space:

1. **First type of connections**: in the upper zone of the quantum, the south-western and north-eastern faces are used, and in the lower zone, the north-western and south-eastern faces (see fig. 2). In this figure, the quantum under consideration is colored white, and the neighboring ones are colored black.
2. **Second type of connections**: in the upper zone of the quantum, the north-western and south-eastern faces are used, and in the lower zone, the north-eastern and south-western faces (see fig. 3). In this figure, the quantum under consideration is colored black, and the neighboring ones are colored white.

Fig. 2

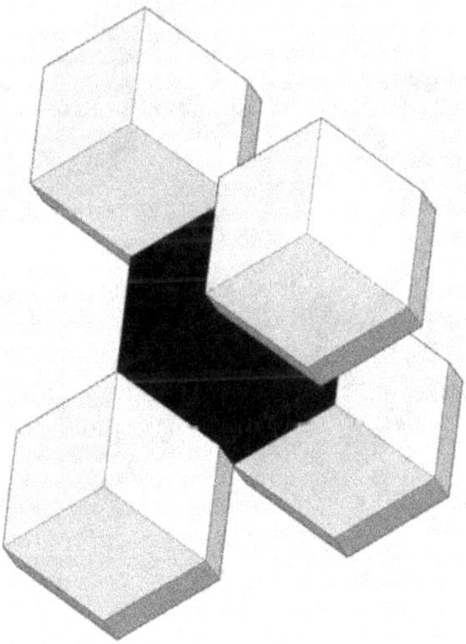

Fig.3

If a quantum of space has the first type of connections, then all four of its neighboring quanta will have the second type of connections, and vice versa. Thus, in accordance with the principle of duality, two half-spaces are formed.

In fig. 4, a fragment of one half-space consisting of 24 quanta of space is shown. The length of the fragment is 3 quanta, the height is 4, and the thickness is 2. The quanta of space from the second half-space are located in the voids between the quanta of the first half-space, as shown in fig. 4. Quanta located at the boundary of space have only two neighboring quanta instead of four.

Fig. 4

PRANA

Clearly, just as a child cannot be conceived without a father, matter cannot emerge without something external to the ether. Therefore, it can be assumed that the **"father" of matter** is that invisible part of the cosmos, which we have called the **cosmic energy field**. For simplicity, we will refer to it as **prana**, which means **life force**. Thus, it can be said that **prana** is the father of matter, **ether** is the mother, and **matter** is the offspring of prana and ether.

POWER

Here, **power** refers to the ability of something to perform a physical action or do work. According to the principle of duality, there are two types of power: **active** (force) and **passive** (energy).

Active power is **prana**. It functions like the sperm cell. Prana is the life-creating force, that is, the **life force**. According to the principle of quantization, prana consists of **quanta**. All prana quanta have the same quantum of force but differ in their directions of action.

Passive power is **the energy of the ether**. The energy of a quantum of ether is a **quantum of energy**. In accordance with the principle of duality, a quantum of energy consists of a dense core and its surrounding field. In one quantum of space, there can only be one core of the quantum of ether. The field of the quantum of ether extends beyond the boundaries of the quantum of space. A quantum of ether functions as the **egg cell**.

Prana and ether are **primary forms of force and energy** that can transform into other forms.

CHAPTER 5. THE BIRTH OF MATTER

QUANTA OF MATTER, MASS AND INERTIA

A stream of life force — prana — is directed at the pra-ether. Some of the prana quanta manage to reach the core of the ether quantum. In this case, the prana quantum strongly compresses the ether quantum, turning it (the core together with the field) into a quantum of matter. Furthermore, the prana quantum transforms the energy of the ether quantum into a quantum of mass. The energy of the prana quantum imparts inertial force to the quantum of mass, equal to the quantum of inertia. The direction of the inertial force coincides with the direction of the prana quantum. This process is analogous to fertilization by a sperm cell.

If in one quantum of space there was only one ether quantum, due to the strong compression of the ether quantum, many quanta of matter may fit into one quantum of space. For the qualitative description of the unified field theory, knowing the maximum number of matter quanta that can fit into one quantum of space is not essential.

MOVEMENT

The force of inertia causes the quantum of matter to move in the direction of this force. However, instead of movement, the states of two neighboring quanta of space change. In the quantum of space where the quantum of matter was located, a quantum of ether appears, and in the neighboring quantum (the one most aligned with the direction of the inertial force), a quantum of matter appears instead of the quantum of ether. However, for simplicity, we will conditionally call this change of state in two neighboring quanta of space the familiar term — movement (more precisely, the quantum of movement). In reality, no movement occurs, just like on the movie screen.

The described movement can occur in any of the half-spaces. If the quanta of matter are formed in two different half-spaces, they cannot collide.

TIME

The change of state of two neighboring quanta of space occurs in a very short period of time, which is called the quantum of time. In nature, no process occurs in less than one quantum of time. This is how movement in space and time appears.

QUANTUM OF DISTANCE

The distance between the centers of two neighboring quanta of space is the minimum distance in nature. It is called the quantum of distance. In one quantum of time, a quantum of matter travels a distance equal to one quantum of distance.

MOVEMENT SPEED

The speed of such movement in the pra-ether is the maximum possible. Indeed, a quantum of matter can only travel one quantum of distance in one quantum of time (not considering degenerate quanta of space, which will be discussed later in the "Field" section of Chapter 6, "Particles"). It also cannot travel a quantum of distance in less than one quantum of time, as this is the minimum possible time.

KINETIC ENERGY

With the emergence of the speed of movement, kinetic energy also appears, which for a quantum of matter depends on its speed of movement. For a particle, kinetic energy depends not only on its speed of movement but also on its mass (see the "Movement of Particles" section of Chapter 6, "Particle Formation").

TRAJECTORIES OF INERTIAL MOTION

Inertial motion of a quantum of matter is not linear, because in the network of quantum spaces, there is no possibility to move in a straight line. The force of inertia gives the quantum of matter a certain direction of movement. However, the quantum of matter can move through each quantum of space only in ten directions. In six of these directions, movement follows wave trajectories, and in four directions, it follows spiral trajectories. Two wave motion trajectories are shown in Figure 5, and four are shown in Figure 6. Two spiral motion trajectories are shown in Figure 7, and two are shown in Figure 8.

For greater clarity in the figures 6–8, the trajectories are depicted only in one direction from the considered quantum of space. In Figures 7–8, the considered quantum of space is colored red. In Figure 7, the right spiral trajectory is shown above the red quantum of space, and the left one below it. In Figure 8, this is reversed.

The quantum of matter moves in the direction from the ten available directions that deviates the least from the prescribed direction. Upon reaching the next quantum of space, the quantum of matter continues its movement along one of the ten possible trajectories. The choice of trajectory is again made based on minimizing the deviation from the prescribed direction of motion. Thus, if the direction of motion does not coincide with any of the ten directions, the trajectory of movement turns out to be a complex line consisting of segments of wave and spiral lines.

If a quantum of matter reaches the boundary of space, its further movement occurs as follows. Since boundary quanta of space only have two neighboring quanta of space, the quantum of matter, entering through one neighboring quantum of space, will be forced to exit through the second neighboring quantum of space. In this process, the direction of the quantum's motion will clearly be disrupted. Striving to restore its direction of movement, the quantum of matter will soon find itself back at the boundary quantum of space. Thus, the quantum of matter will continue its movement along a closed trajectory at the very boundary of space.

Since there is no possibility of moving in a straight line in the network of quantum spaces, the quantum of matter acquires two speeds: along the trajectory of motion (real speed) and along the direction of movement (calculated speed).

Fig. 5

Fig. 6

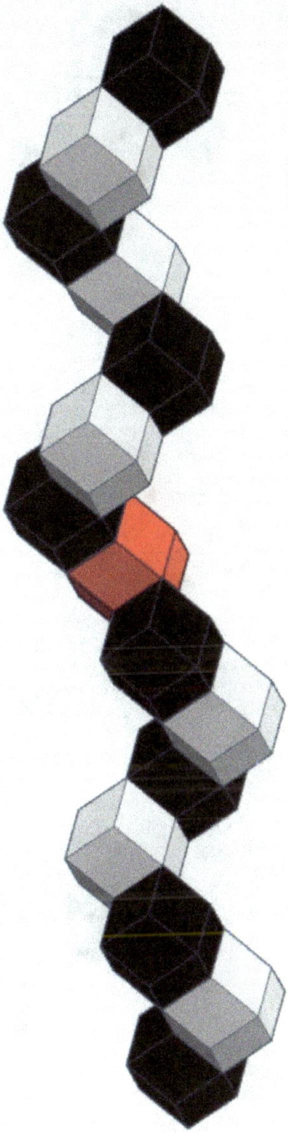

Fig. 7

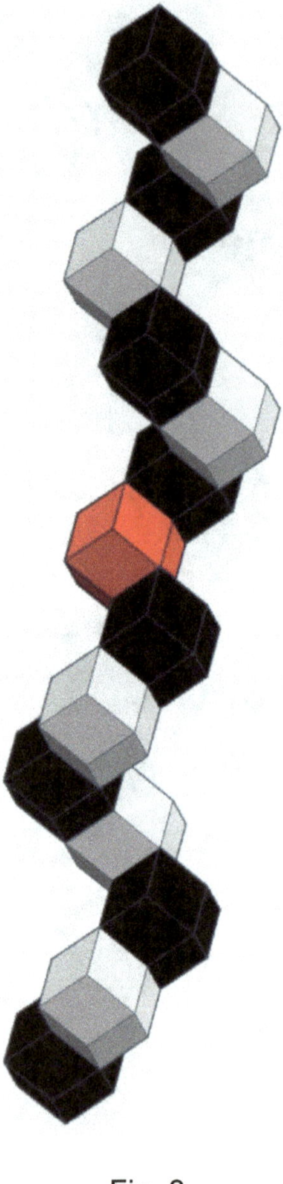

Fig. 8

QUANTUM OF FRICTION

When a quantum of matter moves in the direction of the inertia force, quanta of ether move in the opposite direction. Due to the fields of quanta of ether, the moving quantum of ether encounters a slight resistance from neighboring stationary quanta of ether from the opposite half-space. This process is analogous to friction when bodies move through any medium. The resistance force to the movement of a quantum of ether is the minimal force and is called the quantum of friction. In magnitude, the quantum of friction is equal to the quantum of inertia, but it has the opposite direction to the inertia force. Thus, the frictional force and the inertia force balance each other out. The presence of friction prevents the quantum of matter from moving along its trajectory with infinite speed, so the maximum speed of movement of the quantum of matter, particles, and bodies is limited.

ANISOTROPY OF SPACE

Space cannot be isotropic due to its quantum structure. As for the practical perception of space as isotropic, it is explained as follows:

Since the mass of quanta of matter is constant, the speed of a quantum of matter along its trajectory depends solely on its kinetic energy. During the process of materialization, the quanta of matter receive the same amount of kinetic energy. However, they spend different portions of their energy to overcome inertia when passing a unit of distance along their trajectory. The minimal energy expenditure occurs when the quantum of matter moves along a helical path, as this results in the least change in the velocity vector. The greatest energy expenditure occurs when moving along a wave-like path, as this trajectory causes significant changes in the velocity vector.

Since the quanta of matter spend different amounts of their energy per unit of distance depending on their path, the speed of the quanta along their trajectory, which depends on the remaining energy, turns out to be different (greater for the helical path, lesser for the wave-like trajectory).

Finally, per unit of distance in the direction of movement, the helical trajectory requires traveling a greater distance (but at a higher speed), while the wave-like trajectory requires a shorter distance (but at a lower speed). Thus, in both cases, the time spent to cover a unit of distance in the direction of motion turns out to be equal.

CHAPTER 6. PARTICLE FORMATION

Further growth of matter occurs similarly to the growth of the fetus in the womb.

PARTICLE GROWTH

The quanta of prana impart motion to the quanta of matter in various directions. As a result, a situation may arise where two quanta of matter simultaneously enter the same quantum of space from different sides. In this case, a new particle is formed with doubled mass and inertia. The direction of its movement will be determined by the directions of the quanta of matter that formed this particle. This is how particles are created.

If the motion of the quanta of matter was diametrically opposite, the new particle will move along a closed trajectory. This is due to the fact that a quantum of space does not have diametrically opposite neighboring quanta. Therefore, the quanta of matter, having entered this quantum of space through two of its four neighboring quanta, will inevitably exit through one of the remaining two. In this process, the direction of the particle's motion will be altered. In an attempt to restore its original direction, the particle will soon find itself back in the same quantum of space.

FIELD

Since the quantum of space where two quanta of matter appeared contained only one quantum of ether, this quantum of ether can remain in only one of the two quanta of space from which these quanta of matter came. The other quantum of space will be left without a quantum of ether, making it rarefied. Such a quantum of space is called degenerate.

According to the principle of materiality, a degenerate quantum of space cannot be empty. Although it contains neither a quantum of matter nor a quantum of ether, it holds the fields of ether quanta located in the surrounding quanta of space.

Thus, a degenerate quantum of space appears next to a particle with doubled mass. As this particle moves, it "transitions" to the next quantum of space along its trajectory. In this process, the degenerate quantum of space regains its quantum of ether, while the quantum of space previously occupied by the particle becomes degenerate.

Consequently, a degenerate quantum of space always accompanies a moving particle with doubled mass.

As larger particles form, the number of degenerate quanta of space around the particle increases. This results in the formation of a surrounding field, which includes both quanta of space containing ether quanta and degenerate quanta of space. The greater the mass of the particle, the larger and more symmetrical its field. The concentration of degenerate quanta of space decreases with distance from the particle.

This field has boundaries: the degenerate quanta of space belonging to the particle are located only within the field. The field moves together with the particle as a single entity. A quantum of matter has no field of its own, which makes its detection more difficult.

CELLULAR STRUCTURE OF MATTER

As a particle continues to grow, there inevitably comes a point when the combined quanta of matter can no longer fit within a single quantum of space. At this stage, further particle growth is accompanied by the filling of an adjacent quantum of space with quanta of matter.

Thus, as the particle expands, a cellular structure of matter emerges, where a quantum of space filled with quanta of matter functions as a cell. As the mass of the particle increases, the number of such cells (occupied quanta of space) also grows, similar to the increase in cell numbers in a living organism as it develops. Further mass growth leads to the formation of larger bodies with an even greater number of cells.

Accordingly, in line with the principle of similarity, particles—like living organisms—can be subcellular, unicellular, or multicellular.

It is likely that most subcellular particles consist solely of quanta of matter, just as the simplest atoms (hydrogen atoms) make up approximately 88.6% of all atoms in the universe [8]. Furthermore, quanta of matter may constitute a significant portion of what is known as dark matter [9].

REDUCING SPEED OF MOVEMENT

As long as a particle fits within a single quantum of space, it can transition to an adjacent quantum of space within one quantum of time, meaning it moves by one quantum of distance. However, as the particle grows in size, it requires a second quantum of space. Now, to cover the same distance, it will take two quanta of time, since within a single quantum of time, only the first of the two occupied quanta of space can change its state. As a result of this state change, a quantum of ether will appear in the first quantum of space. Only after this, in another quantum of time, can the second occupied quantum of space switch its state with the first, now containing the quantum of ether.

With further particle growth, a third and subsequent quanta of space become necessary. Thus, as particles and bodies increase in size, their maximum possible speed of movement continues to decrease.

PARTICLE MOVEMENT

The kinetic energy of a particle is determined not only by its mass but also by its velocity. In the proto-ether, a particle that fits within a single quantum of space moves at a constant speed, regardless of direction or mass. However, when moving within a field, the situation changes. Let us examine the motion of a particle under different field movement scenarios.

MOTION OF A PARTICLE IN A STATIONARY FIELD

As a particle approaches another particle (or body), it enters the surrounding field. Unlike the proto-ether, the field contains degenerate quanta of space. As previously mentioned (see Chapter 5, "The Birth of Matter," section "Motion," subsection "Quantum of Friction"), a moving quantum of ether encounters slight resistance from the neighboring stationary ether quanta of another half-space. However, degenerate quanta of space do not experience such resistance.

Therefore, if a moving particle encounters a degenerate quantum of space along its path, its inertia is not counterbalanced by friction (which is absent in this case). This results in an acceleration of the particle's motion. The particle now transitions not just between adjacent quanta of space but between more distant ones. The extent of these jumps depends on the number of degenerate quanta of space along the particle's trajectory. Figuratively speaking, the particle's "steps" from one quantum of space to the next become "jumps" spanning two or more quanta along its trajectory.

Thus, the particle's velocity within the field begins to increase. As it moves closer to the center of the field (i.e., to another particle), its speed continues to rise. If the particle does not merge with the other particle, it will reach its maximum velocity at the closest point of its trajectory to the center of that particle. The greater the mass of the other particle (and consequently, the larger its field), the more rarefied the space experienced by the approaching particle, and the greater its acceleration.

MOVEMENT OF A PARTICLE IN THE MOVING FIELD OF A PARTICLE RECEIVER

When the field moves toward the particle, the relative speed of their motion will be the sum of their individual velocities. However, it should be noted that during the opposing motion of the ether quanta of the field, the length of the particle's "jumps" shortens, which leads to a decrease in its speed. Thus, when the particle reaches another particle (or body), it does not acquire the speed it would have attained in a stationary field.

Therefore, although the velocities of the particle and the field add up during their opposing motion, the relative speed of the particle does not increase significantly due to the decrease in its motion within the opposing field.

In the case of co-moving particles and field, the relative speed of the particle decreases by the speed of the co-moving field, but the particle's speed within the field increases because the length of its "jumps" increases, which compensates for the reduction in relative speed.

Thus, regardless of the direction of the field's movement, the speed at which a particle reaches another particle remains virtually unchanged. In other words, the device measuring this speed (i.e., the observer) will not detect any change in the speed (as confirmed by the Michelson-Morley experiment). Therefore, the claim of the non-existence of the ether is incorrect.

MOTION OF A PARTICLE IN A MOVING FIELD OF THE PARTICLE SOURCE

If the source of particles is in motion, the length of the "jumps" (and the speed of motion) of a particle moving against the direction of the source field decreases. The length of the "jumps" (and the speed of motion) of a particle moving in the same direction as the field of the source increases compared to the length of the "jumps" (and the speed of motion) of a particle in a stationary field of the source.

However, when the particle reaches the boundary of the field, their speeds practically become equal. Further motion of the particles in the proto-ether field occurs with the same speed in all directions, regardless of the direction and speed of the source's motion.

Thus, the addition or subtraction of the velocities of the particles and the source occurs only within the field. Outside the field of the source, the velocity of the particles' motion is independent of the speed and direction of their source. The claim about the absolute independence of the speed of light (photons) from the direction and speed of the source's motion (including within the source's field) is incorrect.

INTERACTION

The essence of the unified field theory is that every interaction must have one common cause. And that cause is the pra-ether. Why is the pra-ether the only cause of any interaction?

THE NATURE OF GRAVITY

In the case of a large distance between particles, the symmetric field of the particle is almost in equilibrium with the surrounding pra-ether. However, as the particle moves, a situation inevitably arises where its field intersects with the field of another particle. In such a case, the equilibrium of the particle with the pra-ether is disturbed. The sparse field of the other particle will begin to act on it, while the pra-ether will continue to act from the opposite side. Thus, two forces arise, and the pra-ether, acting on each of these particles, draws them closer to one another.

This creates the impression of attraction between the particles, although in reality, the particles do not directly interact with each other. This phenomenon explains the nature of gravity and is known to science as the law of universal gravitation, discovered by Isaac Newton in 1687. It would be more accurate to call it the law of universal convergence, as there is no actual pulling or attraction between particles (or bodies). However, the convergence of particles due to the influence of the pra-ether will be referred to by the familiar term—gravity.

Unfortunately, the formula for determining the force of gravity has proven to be valid only for the macroworld. The smaller the distance between particles, the more the interaction between them deviates from the law of universal gravitation, so this law cannot claim universality.

If we calculate the forces with which the pra-ether acts on each of the two particles, the values of these forces will be accurate not only for the macroworld (bodies) but also for the microworld (particles). It is even possible that these two forces could be different. According to these formulas, the forces acting on particles will be significantly higher than those obtained using Newton's formula. For charged particles, they will match the values obtained using Coulomb's law, and for protons within the atomic nucleus, they will match the nuclear forces.

An attempt to derive these formulas was made in [5]. However, this derivation was based on the assumed distribution of ether density in the particle's (body's) field. To identify a more accurate distribution of ether density, it is necessary to know how many quanta of matter fit into one quantum of space.

Thus, gravity is not explained by the presence of a gravitational field but by the influence of the pra-ether on particles or bodies that have come closer to each other. Since there is no gravitational field, there can also be no gravitons, which physicists have unsuccessfully tried to detect for almost a century.

STRUCTURE

Gravitational interactions between particles, following the principle of organization, can lead to the formation of a new particle with a structure distinct from the original interacting particles.

New structures emerge through one of four mechanisms:

1. **Fusion of particles** within the same half-space, forming a new particle whose mass equals the sum of the interacting particles' masses.
2. **Formation of a symmetric structure** consisting of two identical particles located in different half-spaces (e.g., a stable electrically neutral particle or a dipole).
3. **Formation of an asymmetric structure** consisting of two different particles located in different half-spaces (e.g., an unstable electrically neutral particle).
4. **Planetary structure**, where a smaller-mass particle orbits a larger-mass particle across two half-spaces (e.g., a light hydrogen atom). These structures occupy significantly more space than the first three types.

The formation of particle structures is significantly influenced by the dual nature of space.

DUAL STRUCTURE OF SPACE

As stated above, space consists of two half-spaces (see the section 'Space' in the paragraph 'Ether' of Chapter 4 'Pra-ether'). "As long as the particles are small, the dual structure of space has little effect on their interaction. The field of a particle composed of one or several quanta of matter is limited, containing only a small number of ether quanta. However, as the particle grows, its field expands, incorporating more ether quanta. At a certain point, some ether quanta from the opposite half-space become surrounded by multiple ether quanta from the moving particle's field and are drawn along with it. This occurs due to the inherent fields present in ether quanta.

All known elementary particles in science are structures of varying complexity and, strictly speaking, are not truly elementary. The only genuinely elementary entities are particles composed of one or a few quanta of matter. A particle made up of two elementary particles already constitutes a structure. Thus, even dark matter, which primarily consists of quanta of matter and particles smaller than known elementary particles, possesses its own structures.

Let us examine some of the most notable structures formed by particles.

ANTIPARTICLE

An antiparticle differs from a particle only in that it is located in a different half-space. For example, the electron is located in the negative half-space, while the positron is in the positive half-space. Similarly, the proton is located in the positive half-space, while the antiproton is in the negative half-space. If one part of a particle with mass m1 is located in the negative half-space and the other part with mass m2 is in the positive half-space (for example, a neutron), then its antiparticle (the antineutron) will have mass m1 in the positive half-space and mass m2 in the negative half-space. If a particle consists of two identical particles located in different half-spaces, then such a particle is called truly neutral, and it does not differ from its antiparticle (for example, a photon). Thus, a particle differs from its antiparticle only by its location in different half-spaces.

NEUTRAL PARTICLE

The most common neutral particle is the photon, which in science is defined as a fundamental, truly neutral particle. However, according to the principles of the unified field theory being presented, the fundamental particle is actually the quantum of matter, which constitutes most of dark matter. Therefore, the photon is a composite particle, made up of two identical particles of dark matter.

When particles with equal mass are formed simultaneously in two neighboring quanta of space, belonging to different half-spaces, the fields of each particle will have their centers in adjacent quanta of space (i.e., they will practically coincide). In this case, these particles can be considered as a single particle with double the mass and a common field. Such particles are known to science as stable electrically neutral particles (for example, the photon).

However, if the masses of the particles in different half-spaces are different, for example, m1 and m2 (m1 > m2), then such neutral particles with a mass equal to the sum of m1 and m2 will be unstable (for example, the neutron).

CHARGED PARTICLE

If a particle is formed in only one of the two half-spaces (for example, in the positive one), then its field will be created only in the positive half-space. As mentioned earlier (see the section "Dual Structure of Space"), such a particle, in its motion, will drag not only its own field but also a part of the other (negative) half-space. This part is located within the particle's field. Inside it are stationary ether quanta, which are set in motion by the moving ether quanta from the positive half-space. However, nothing visible occurs in the negative half-space because there is an "exchange" of ether quanta between adjacent interacting space quanta. Still, this process does occur.

Examples of the formation of "charged" particles are: in the positive half-space – the proton, and in the negative half-space – the electron. The word "charged" is put in quotation marks because neither the proton nor the electron has a charge. The apparent existence of charges can be explained as follows:

Consider two cases of particle approach. In the first case, both particles are located in the same half-space. After their fields intersect, the quanta of space from one particle's field will "exchange" their content with the quanta of space from the other particle's field. In the other half-space, the ether quanta dragged by both particles will simultaneously try to occupy the same space. But this is impossible, because only one ether quantum can fit in a single quantum of space.

This picture resembles a situation where two elastic bodies in contact are compressed, leading to their elastic deformation and the formation of a reactive force that seeks to separate them. Something similar occurs with ether quanta from the other half-space, dragged by two particles that are approaching each other. As a result, these ether quanta push against each other, being unable to occupy the same space to which they are both attempting to move. Thus, a large reactive force arises, corresponding to the repulsive force in Coulomb's law for like electrical charges.

A similar picture arises when an unstable neutral particle (for example, a neutron) with masses m1 and m2, located in different half-spaces, approaches another particle with mass m, which is located in the half-space containing most of the mass (i.e., m1) of the unstable neutral particle. However, in this case, the magnitude of the reactive force will be smaller than it would be for the approach of two particles with masses m1 and m, both located in the same half-space. This is because the ether quanta of the field dragged by the particle with mass m will only occasionally try to occupy the same quantum of space as the ether quanta of the field created by the mass m2 of the unstable neutral particle.

In the case of two particles approaching each other, located in different half-spaces, after their fields intersect, the ether quanta of one particle's field will "exchange" their content with the ether quanta dragged by the other particle. The reactive force that arose in the first case will not appear here. This is because the ether quanta dragged by the moving particles are in different half-spaces, not in the same half-space as in the first case.

Furthermore, the ether quanta dragged by the motion of one particle move towards the other particle, which accelerates their movement. This can be explained as follows: the ether quanta dragged by the movement of a particle from the other half-space, at the moment they change their space quanta (i.e., after one quantum of time), bind together into a whole. At this moment, the other particle attempts to move into the quantum of space occupied by one of the two bound ether quanta. Therefore, the particle exchanges its space quantum with the two bound ether quanta and moves by two space quanta.

The particles continue to move toward each other until they form a neutral particle, which will then move as one particle. Its velocity vector will depend on the velocity vectors of the merging particles. The described interaction between the two particles reflects Coulomb's law for the interaction of oppositely charged electrical charges.

Thus, the attraction of oppositely charged and the repulsion of like charges are nothing more than the interaction of particles in different or the same half-space. In other words, electric charges do not physically exist. The electric charge is a fundamental physical concept that cannot logically be explicitly defined through other simpler concepts. The existence of the electric charge cannot be proven by any single experiment; it follows from the entire set of known experimental facts of electrodynamics [10].

DIPOLE

As noted earlier (see the "Structure" section), one of the four possible ways of forming structures is through the creation of a special stable particle. This particle, known in science as a magnetic dipole, consists of two small elementary (or even the simplest) particles located in close proximity to each other but in different half-spaces.

The peculiarity of a dipole is that it can combine with similar dipoles either in parallel or in sequence, forming a larger stable body known as a magnet. The interaction of a magnet with individual magnetic dipoles creates the illusion of magnetic field lines, which do not actually exist.

Similarly, an electric dipole is formed, but it consists of larger elementary particles, such as an electron and an ion.

ATOMIC NUCLEUS

The nature of electric charges described above (see the section "Charged Particle") also explains the nature of so-called nuclear forces. According to science, in the case of strong nuclear interactions, nuclear forces hold the positively charged protons in the nucleus of an atom. In reality, there are no repelling like electric charges, but only ether quanta pushing against each other. These ether quanta, despite the forces trying to bring them together, cannot occupy the same space. Therefore, there is no need to invoke mysterious nuclear forces that supposedly hold like charges apart, namely protons, in the nucleus of the atom. The force that tends to separate them is simply a reactive force caused by the action of external active gravitational forces on the ether quanta pushing against each other.

In weak nuclear interactions, according to science, nuclear forces lead to beta decay. As a result of this decay, a neutron turns into a proton. In fact, during the formation of the unstable particle (neutron), the particle with mass m1 approached a particle with a smaller mass m2, located in another half-space. The fields of these particles had different sizes. At the first stage of the approach, from the moment the fields touched until half of the smaller field was embedded into the larger field, the force acting on the smaller particle increased, then began to decrease until it disappeared entirely. The accompanying resistance force, however, kept increasing. This imbalance of forces initially caused the particle with mass m2 to stop and then move backward, which is perceived as the decay of the neutron into a proton and electron.

The nucleus of an atom consists of protons and neutrons. However, the number of neutrons can vary with the same number of protons. This leads to a wide variety of nuclei, many of which turn out to be unstable. An excessive increase in the mass of nuclei also leads to their instability. Such unstable nuclei make up more than 90% of the total number of nuclei. An unstable nucleus gradually decays into constituent parts, which may result in the formation of a stable nucleus.

ATOM

As mentioned earlier (see the section "Structure"), one of the two ways of forming a new structure is the creation of a planetary-type structure. In this case, a particle with a smaller mass begins to rotate around a particle with a larger mass. This forms a planetary-type structure, which occupies a large space in the universe.

One such stable structure is the atom, consisting of a nucleus and one or more electrons rotating around it. The nucleus, made up of one or more protons, is located in the positive half-space, while the electrons are in the negative half-space. The nucleus of the atom may also contain neutrons, which are partially located in the positive and partially in the negative half-space.

If the nucleus of the atom is formed in the negative half-space, and the particles rotating around the nucleus are in the positive half-space, then such an atom is an antiatom. When an antiatom meets an atom of matter, their structures mutually destroy each other, forming small particles.

Thus, antimatter is no different from matter, except for its location in a different half-space. Since, during the formation of structures, there was more matter than antimatter, new antimatter formations are quickly destroyed by matter. This is why it is so difficult to detect antimatter in our space.

MOLECULE

There are particles that, after combining, create a stable structure. Let's call them conjugated particles. Examples of conjugated particles include protons and neutrons, which form the nucleus of an atom. Another example is the nucleus and the electron, which form the atom.

Molecules are much more complex structures than atoms. They form due to conjugation, such as the conjugation of a hydrogen atom with atoms of more complex elements. This forms molecules of various hydrogen compounds. Another atom that conjugates with atoms of complex elements is the oxygen atom. The combination of these atoms leads to the formation of molecules of various oxygen compounds.

This is how simple and very complex molecules of numerous inorganic and organic substances are formed.

CHAPTER 7. BIRTH, DEVELOPMENT AND DEATH OF SPACE

The cosmos contains a vast variety of celestial bodies, formed in many different ways. However, initially, they were formed by the method described below. Subsequently, through their interactions, collisions, and decay, new celestial bodies began to form in entirely different ways.

In accordance with the principle of similarity, the birth, development, and death of celestial bodies occur in a manner analogous to the birth, development, and death of humans.

BIRTH OF SPACE BODIES

Under the influence of gravitational forces, particles (elementary particles, atomic nuclei, atoms, and molecules) come closer together, forming something resembling a rarefied gaseous cloud. Over time, this cloud becomes increasingly compressed. Eventually, it forms a gaseous celestial body in the shape of a sphere, primarily composed of hydrogen atoms.

Further subjected to gravitational forces, this first celestial body grows by accumulating particles and bodies from its surroundings. The growth of the body leads to its compression. Finally, due to the increased pressure at the center of this sphere, liquid droplets begin to form. These droplets grow in size and, merging with one another, create a liquid core inside the gaseous celestial body. This is how liquid celestial bodies with a gas shell are born. Around these liquid celestial bodies, the first significant fields, in cosmic scale, form with degenerate space quanta.

DEVELOPMENT OF SPACE BODIES

When celestial bodies come close to one another, and the influence of gravitational forces begins to manifest, they merge into larger bodies. At the center of such large bodies, high pressure develops. This pressure promotes the formation of larger atoms of various elements. The growth in the number of large atoms leads to the formation of a solid core at the center of the celestial body. The gradual growth of the solid core results in the formation of a solid celestial body, with liquid masses still remaining on its surface. These solid-liquid celestial bodies continue to maintain a gaseous shell – an atmosphere – around them. This is how the first planets, resembling Earth, are formed.

Under the influence of gravitational forces, moving solid celestial bodies merge with each other and absorb less massive gaseous and liquid celestial bodies. This accelerates their growth and increases the pressure at the center of the body. Excessive pressure in their center leads to the destruction of the electron shells of atoms, releasing enormous amounts of thermal energy. This is how plasma cores form in solid celestial bodies, similar to Earth.

The further growth of planets, accompanied by an increase in internal pressure, leads to the fusion of nuclei in the planet's plasma core. When the nuclei fuse, even more energy is released than when the electron shells of atoms are destroyed. This energy heats the planet so much that it begins to emit tiny elementary particles into the surrounding space, including photons that create light. This is how the first stars are formed.

The process described above is accompanied by the growth of a field around the celestial body (planets, stars, and star systems).

AGING OF SPACE BODIES

The further growth of the mass of stars, due to mergers with other celestial bodies, leads to the formation of supermassive stars with excessive internal pressure and an external field of enormous size.

This last factor promotes continued growth of the star's mass by absorbing other celestial bodies that are at a sufficiently large distance from it. Eventually, a point is reached when the photons emitted by the star are unable to overcome the gravitational pull of the star. This causes the star to stop emitting light into space and become invisible. This is how black holes are formed.

This process is accompanied by the further growth of the field around supermassive celestial bodies (giant stars and black holes).

DEATH OF SPACE

In accordance with the principle of similarity, for stars, galaxies, and the entire cosmos, there comes a moment similar to the death of a human being.

Supermassive stars (especially black holes) are unstable celestial bodies because their enormous mass creates fields of such size that a huge mass of cosmic matter falls into their sphere. Under the influence of gravity, this entire mass of cosmic matter eventually merges with the supermassive star. This results in internal pressure that destroys the atomic nuclei at the center of the star. When atomic nuclei are destroyed, a large amount of energy is released, causing the star to explode. As a result of the explosion, a significant mass of the star's substance is ejected into space. Thus, the life cycle of a massive unstable celestial body concludes, as it can only leave behind a small body.

The destruction of galaxies occurs as follows. Due to the limitations of space, no matter in which direction a galaxy moves, it will eventually reach the boundary of space. Over time, this creates the impression that all galaxies are flying apart from some kind of explosion, which, in reality, never occurred. In the case of an explosion, the galaxies' separation would slow down rather than accelerate, as is happening in reality. This acceleration is explained as follows.

As the gigantic fields of galaxies approach the boundaries of space, the equilibrium of the galaxies' fields is disturbed. From the side of the boundary of space, the galaxies' fields do not experience the influence of the pra-ether that they continue to experience from the opposite side. This leads to the appearance of a force directed toward the boundary of space. This force accelerates the galaxies' movement toward the boundary of space. Thus, the boundaries of space act like a magnet, attracting galaxies that approach them. The accelerated movement of galaxies toward the boundary of space, coupled with the inability to move further, leads to the destruction of the galaxies.

Eventually, the cosmos will become one vast field with pra-ether at its center and matter at its boundaries. But all matter will be deprived of the ability to move due to the complete absence of pra-ether at the boundaries. This marks the death of the cosmos.

FINDINGS

The conclusions derived from the foundations of the Unified Field Theory presented above significantly differ from modern physics. On one hand, they introduce many concepts that are not yet accepted by science for various reasons. On the other hand, they provide either alternative interpretations of well-known physical facts or explanations for phenomena that remain unexplained by modern science. Finally, some of these conclusions fundamentally contradict established scientific theories.

New Concepts Not Present in Modern Physics:

- **Prana** — the energy field of the cosmos;
- **Ether** — the precursor to matter (pra-matter);
- **Pra-ether** — the initial form of ether;
- **Quantization of space** — its fundamental property;
- **Space quantum** — the smallest unit of space, shaped like a rhombic dodecahedron;
- **Quantum of distance** — the minimum possible separation between objects;
- **Quantum of prana** — the smallest portion of prana;
- **Quantum of force** — the force of a prana quantum;
- **Quantum of ether** — the smallest portion of ether contained in a space quantum;
- **Quantum of energy** — the energy of an ether quantum;

- **Quantum of matter** — the smallest particle of matter;
- **Quantum of mass** — the mass of a matter quantum;
- **Quantum of inertia** — the inertia of a mass quantum;
- **Quantum of time** — the smallest time interval required for the state transition of adjacent non-degenerate space quanta;
- **Quantum of friction** — the minimal force opposing motion;
- **Cellular structure of matter** — the fundamental principle of matter organization;
- **Material cell** — a space quantum filled with matter quanta;
- **Two semi-spaces** — the components of the structure of real space;
- **Limited field with degenerate space quanta around all objects** — the sole cause of any interaction.

New Interpretations of Modern Physics Principles:

- Motion is not continuous, but consists of a sequential change of states of two adjacent non-degenerate quanta of semi-space.
- The limitation of maximum speed is explained not as an axiom, but by the presence of friction in the ether.
- The dependence of maximum speed on the size of the object is explained not by the increase of mass with speed, but by the cellular structure of matter.

- The expansion of the Universe is not caused by the Big Bang but by the finite nature of space, which inevitably leads to a situation where all galaxies move from the center toward its boundaries.
- The accelerated expansion of the universe is explained not by dark energy, but by the imbalance of forces acting on objects as they approach the boundaries of space.
- The interaction of objects in the macroworld is caused not by gravity (gravitons), but by the imbalance of forces acting on an object within the field of another object.
- The repulsion of like charges is explained not by electric charges, but by the inability of ether quanta, moving in opposite directions, to occupy the same place in a different semi-space.
- The attraction of so-called opposite charges is not due to the presence of electric charges but rather to the acceleration of a particle's motion caused by the counterflow of ether quanta, which is induced by the movement of a particle in the opposite half-space.
- The reason for the retention of protons close together in the atomic nucleus is not the so-called nuclear force, but the force that draws protons together exceeding the force resisting their merging.
- Beta decay in atomic nuclei is not caused by the so-called weak interaction, but by the imbalance of forces (attraction and resistance) that arise during the formation of an unstable neutral particle (neutron).

- Magnetism is not explained by the existence of a magnetic field but by special stable particles—magnetic dipoles—that can combine with each other both in parallel and in series.
- Acceleration of objects as they approach is not explained by the law of gravity but by the transformation of "steps" into "jumps" as an object moves within another's field.
- The independence of the speed of light from the observer's motion is due not to a violation of Galileo's principle of relativity but to a change in the particle's velocity in a moving field, which manifests as a change in the length of its "jumps".

Explanations for Phenomena Not Yet Addressed by Modern Physics:

- Dark energy is **pra-ether**.
- Dark matter consists of the smallest particles, from matter quanta to known elementary particles.
- Galaxies without dark matter are explained by the fact that, during the coalescence of particles, the amount of dark matter decreases and may even disappear entirely.
- Empty areas of space that perplex scientists are explained by the existence of non-materialized regions of pra-ether in modern space.

Key Contradictions with Modern Science:

- Space is anisotropic due to its quantization, but it is perceived as isotropic because of the difference in speeds along different trajectories.
- The actual trajectory of inertial motion consists of wave-like and helical lines. However, it is perceived as linear, circular, or other simple forms due to the small amplitude of the wave component and the diameter of the helical trajectory.
- Galileo's principle of relativity holds even at the limit of speeds, despite the appearance of its violation, due to the change in speed when moving in a moving field.
- The limitation of the area in which objects can interact is due to the limited size of their fields, but this is perceived as unlimited because of the vastness of the fields of cosmic bodies.

Thus, many of the fundamental principles of modern science prove to be false or illusory. This shows that the vast benefits of scientific knowledge do not guarantee its truth. Examples of this include almost all scientific theories, from Ptolemaic astronomy to Einstein's theory of relativity.

Einstein's genius lay not so much in creating the theory of relativity as in realizing the existence of a single cause of all interactions in nature. He was so convinced of this that he dedicated the last thirty years of his life to attempting to develop a unified field theory.

CONCLUSION

The fundamental principles of the Unified Field Theory outlined above have established:

- the existence of invisible precursors of matter – prana and ether;
- the finiteness of the cosmos in space and time;
- the quantum structure of all fundamental concepts without exception: prana, force, ether, energy, space, distance, matter, mass, inertia, motion, time, friction;
- the nature of motion;
- the anisotropy of space and the reason for its apparent isotropy;
- the presence of friction in ether (explaining the limitation of maximum motion speed);
- the cellular structure of matter, which causes the deceleration of motion with increasing mass;
- the nature of the field surrounding material objects, which serves as the sole indirect cause of all interactions both in the micro and macro world;
- the nature of gravity, electric charges, nuclear forces, and magnetism;
- the reason for the invariance of the speed of light regardless of the movement of either the source or the receiver.

The foundations of the Unified Field Theory have made it possible to explain all known types of interactions (gravitational, electromagnetic, strong and weak nuclear) through a single cause and to present a perspective on the origin of the cosmos different from the Big Bang theory. Moreover, they have provided explanations for the following scientific discoveries, which remain unexplained by modern physics: dark matter (1922), the expansion of the Universe (1929), the accelerated expansion of the Universe (1998), a matter-free region of space (2007), and a galaxy without dark matter (2018).

This research offers only fundamental principles and does not include the mathematical framework necessary for establishing a quantitative description of the Unified Field Theory. Therefore, it is not yet complete. Rather, it serves as an invitation for philosophers and physicists to participate in the continuation of this work. One can hope that, through the joint efforts of philosophers and physicists who join this research, the dream of A. Einstein, W. Heisenberg, and their followers—who dedicated much effort and time to the creation of the Unified Field Theory—will be realized in the near future.

REFERENCES

1. Kika. A new form of knowledge (Moscow, 2020).

2. Unified field theory – Wikipedia.

3. String theory – article on the Internet.

4. The Theory of Everything – Wikipedia.

5. Kika. The Universal Interaction Hypothesis. On the issue of creating a unified field theory (Moscow, 2020).

6. Dark energy – Wikipedia.

7. Astronomers have discovered a “great nothingness” – article on the Internet.

8. Hydrogen – Wikipedia.

9. Dark matter – Wikipedia.

10. Electrostatics: elements of educational physics – article on the Internet.

OTHER BOOKS BY THE AUTHOR

1. Esoterics. Digest of articles. St. Petersburg, 2017.

2. Truth and fate. Collection of poems. Moscow, 2019.

3. Man and his universe (2nd ed.). Amazon, 2020.

4. The path to immortality. A short guide to questions and answers. Moscow, 2020.

5. A new form of knowledge. Moscow, 2020.

6. Hypothesis of universal interaction (on the question of the creation of a unified field theory). New York, 2020.

7. Hypothesis of Universal Interaction. On the issue of creating a unified field theory. Moscow, 2020.

8. The path to justice. Moscow, 2020.

9. Who is a person. Moscow, 2020.

10. What is what. The nature of spirit, man and matter. Moscow, 2021.

11. The truth is within us. Collection of poems and quatrains (3rd ed.). Moscow, 2022.

12. The path to health and happiness. Moscow, 2022.

13. Imaginary world. Amazon, 2024.

REQUEST TO READERS

IF YOU LIKED THIS BOOK, AND FOUND IT USEFUL OR NOT, I WOULD VERY APPRECIATE IF YOU POST A BRIEF REVIEW ON AMAZON. I READ ALL THE REVIEWS PERSONALLY TO CONSTANTLY WRITE WHAT PEOPLE WANT. THANK YOU FOR YOUR SUPPORT!

www.ingramcontent.com/pod-product-compliance
Lightning Source LLC
LaVergne TN
LVHW022315200926
848481LV00004B/1356